How to be Brilliant at

Living Things

Colin Hughes and Winnie Wade

Brilliant Publications

Published by Brilliant Publications,
The Old School Yard,
Leighton Road, Northall,
Dunstable,
Bedfordshire,
LU6 2HA.

Tel: 01525 222844
Fax: 01525 221250
E-mail: sales@brilliantpublications.co.uk
Website: www.brilliantpublications.co.uk

How to be Brilliant at Living Things
Written by Colin Hughes and Winnie Wade
Illustrated by Bridget Dowty

Printed in Malta by Interprint Ltd.

Colin Hughes and Winnie Wade
ISBN 1 897675 666

First published in 2000
10 9 8 7 6 5 4 3 2 1

Other books in this **How to be Brilliant...** series include:

Maths titles
How to be Brilliant at Algebra
How to be Brilliant at Mental Arithmetic
How to be Brilliant at Numbers
How to be Brilliant at Shape and Space
How to be Brilliant at Using a Calculator

English titles
How to be Brilliant at Grammar
How to be Brilliant at Making Books
How to be Brilliant at Reading
How to be Brilliant at Spelling
How to be Brilliant at
        Word Puzzles (due Nov. 2000)
How to be Brilliant at Writing Poetry
How to be Brilliant at Writing Stories

Science titles
How to be Brilliant at Electricity, Light and Sound
How to be Brilliant at Materials
How to be Brilliant at Recording in Science
How to be Brilliant at Science Investigations

Other titles
How to be Brilliant at Christmas Time
How to be Brilliant at Recording in Geography
How to be Brilliant at Recording in History

Contents

Introduction

How to be Brilliant at Living Things contains 35 photocopiable sheets for use with children working at levels 2–5 of the revised National Curriculum (2000) and Scottish levels C–E. The activities are designed to help children develop scientific understanding of the fascinating topics associated with living things. They can be used whenever the need arises for particular activities to support and supplement your existing schemes of work for science. The activities provide learning experiences which can be tailored to meet individual children's needs.

The activities are addressed directly to the children. They are self-contained and many children will be able to work with little additional help from you. You may have some pupils, however, who have the necessary scientific skills and concepts but require your help in reading the sheets.

The children should be encouraged to use the sheets for all aspects of communicating their work. Those activities which require more than a pen or pencil may be completed with basic classroom science resources. These are listed under the **What you need** heading on the sheets. Some of the sheets require the use of an additional resource sheet. Where this is the case, it has been indicated in the text along with the page number.

Links to the National Curriculum
How to be Brilliant at Living Things relates directly to themes 1, 2, 3, 4 and 5 of the Key Stage 2 programme of study for Life Processes and Living Things. The contents page indexes each activity directly to the programme of study, while pages 5 and 6 give details of the programme of study covered in the book.

Links to the National Curriculum

How to be Brilliant at Living Things supports the following elements of the Key Stage 2 Sc 2 Life Processes and Living Things programme of study.

Pupils should be taught:

1   Life processes

a   that the life processes common to humans and other animals include nutrition, movement, growth and reproduction
b   that the life processes common to plants include growth, nutrition and reproduction
c   to make links between life processes in familiar animals and plants and the environments in which they are found.

2   Humans and other animals

Nutrition
a   about the functions and care of teeth
b   about the need for food for activity and growth, and about the importance of an adequate and varied diet for health

Circulation
c   that the heart acts as a pump to circulate the blood through vessels around the body, including through the lungs
d   about the effect of exercise and rest on pulse rate

Movement
e   that humans and some other animals have skeletons and muscles to support and protect their bodies and to help them to move

Growth and reproduction
f   about the main stages of the human life cycle

Health
g   about the effects on the human body of tobacco, alcohol and other drugs, and how these relate to their personal health
h   about the importance of exercise for good health.

3   Green plants

Growth and nutrition
a   the effect of light, air, water and temperature on plant growth
b   the role of the leaf in producing new material for growth
c   that the root anchors the plant, and that water and minerals are taken in through the root and transported through the stem to other parts of the plant

Reproduction
d   about the parts of the flower (for example, stigma, stamen, petal, sepal) and their role in the life cycle of flowering plants, including pollination, seed formation, seed dispersal and germination.

4   Variation and classification

a   to make and use keys
b   how locally occurring animals and plants can be identified and assigned to groups
c   that the variety of plants and animals makes it important to identify them and assign them to groups.

5   Living things in their environment

a   about ways in which living things and the environment need protection

Adaptation
b   about the different plants and animals found in different habitats
c   how animals and plants in two different habitats are suited to their environment

Feeding relationships
d   to use food chains to show feeding relationships in a habitat
e   about how nearly all food chains start with a green plant

Micro-organisms
f   that micro-organisms are living organisms that are often too small to be seen, and that they may be beneficial (for example, in the breakdown of waste, in making bread) or harmful(for example, in causing disease, in causing food to go mouldy).

In addition *How to be Brilliant at Living Things* supports the following elements of the Key Stage 2 Sc 1 Scientific Enquiry programme of study.

Pupils should be taught:

1   Ideas and evidence in science

a   that science is about thinking creatively to try to explain how living and non–living things work, and to establish links between causes and effects (for example,–Jenner's vaccination work)
b   that it is important to test ideas using evidence from observation and measurement

2   Investigative skills

Planning
a   ask questions that can be investigated scientifically and decide how to find answers
b   consider what sources of information, including first-hand experience and a range of other sources, they will use to answer questions
c   think about what might happen or try things out when deciding what to do, what kind of evidence to collect, and what equipment and materials to use
d   make a fair test or comparisons by changing one factor and observing or measuring the effect effect while keeping other factors the same

Obtaining and presenting evidence
e   use simple equipment and materials appropriately and take action to control risks
f   make systematic observations and measurements, including the use of ICT for datalogging
g   check observations and measurements by repeating them where appropriate
h   use a wide range of methods, including diagrams, drawings, tables, bar charts, line graphs and ICT, to communicate data in an appropriate and systematic manner

Considering evidence and evaluating
i   make comparisons and identify simple patterns or associations in their own observations and measurements or other data
j   use observations, measurements or other data to draw conclusions
k   decide whether these conclusions agree with any prediction made and/or whether they enable further predictions to be made
l   use their scientific knowledge and understanding to explain observations, measurements or other data or conclusions
m   review their work and the work of others and describe its significance and limitations.

Life processes we carry out

All living things, whether they are **plants** or **animals**, are able to carry out **life processes**. Living things, including humans can carry out seven life processes.

Listed below are the names of the seven processes that humans carry out. Also below are seven diagrams showing these processes.

■ Draw a line from the diagrams to the names to match them up correctly.

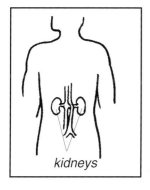
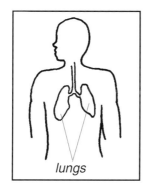
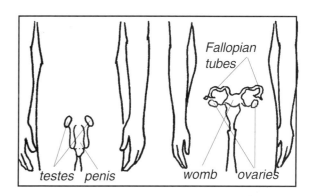

reproduction   nutrition and feeding   excretion

breathing and respiration   sensitivity   growth   movement

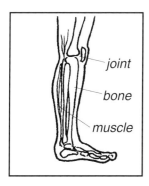
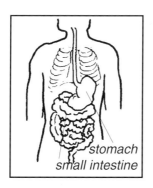

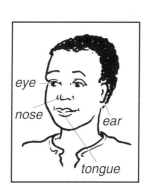

EXTRA!
Draw a table which shows the life process in one column and the organs
in the body which carry out the process in the other.

How to be Brilliant at Living Things

Living things do these!

All living things can do a number of activities or life processes. Living things are also called **organisms**.

- Tick the four boxes which show the main things which **all animals do**.

  feed ☐ grow ☐ hear ☐ hunt ☐

  make a sound ☐ move ☐ reproduce ☐ walk ☐

- **Plants** make their **own food**. Which **two** other life processes do **all** plants do? Tick them.

  make food ✔ grow ☐ hear ☐ hunt ☐

  made a sound ☐ move ☐ reproduce ☐ walk ☐

- Using the above information, finish these sentences:

  A car is not living because ...

  A dried flower is not living because ...

EXTRA!
There are a few more life processes that animals and plants can do.
Can you think of them? If not, find out in a book or CD-ROM.

Life cycles and growth

A cycle is something which goes around. A life cycle shows how an animal or plant reproduces and grows and reproduces again. Some animals change greatly in appearance as they grow. Frogs do and so do butterflies. Another example is dragonflies which are found close to water.

■ Label the diagrams below using these words:

dragonfly emerging
eggs
butterfly
tadpole with legs

frog
tadpole
dragonfly
eggs (frogspawn)

dragonfly nymph
larva (caterpillar)
pupa
eggs

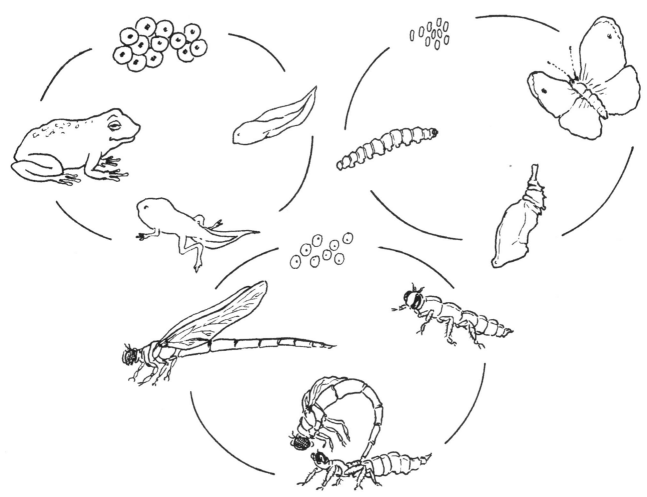

■ Complete the arrows on the three life cycles to show the direction of the cycle.

■ How is the life cycle and growth of humans different from the three animals, above?

EXTRA!
Find out what is meant by the word *metamorphosis*.

Life processes and river animals

The picture below illustrates life in a river. All of these animals carry out
life processes which help them to grow and reproduce to maintain the life cycle.

■ Label the pictures using as many scientific words as you can.

■ Which **four** life processes can be seen in the pictureabove?

1 _____   2 _____

3 _____   4 _____

■ Draw a table to show:
   a   which part or parts of the body help each animal to push through
        the water
   b   the food the animals eat
   c   whether eggs or live young are produced by the female adult.

EXTRA!
Animals hunt their prey using their senses.
Which senses do these animals use?

Teeth – what do they do?

You have four different types of teeth in your mouth.
You need different teeth to do different things.

What you need: crisps or thin piece of celery or peeled carrot.

Look at the drawings to see where the teeth are in your mouth.

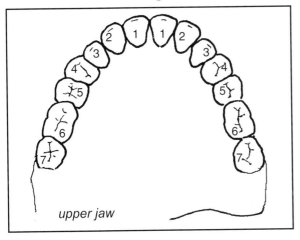

upper jaw

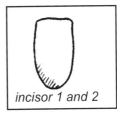

incisor 1 and 2

canine 3

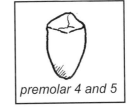

premolar 4 and 5

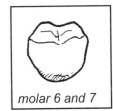

molar 6 and 7

Ask your teacher for a piece of food. Cut, bite, tear, chew and grind the food.

■   Which tooth type are you using to:

chew and grind?_____   cut and bite? _____

tear and grind? _____   tear? _____

Look at these bar charts.

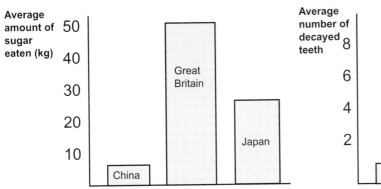

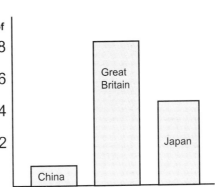

■   Which country has the least tooth decay?
■   In which country do people eat the least sugar?
■   How could you reduce the amount of sugar you eat?

EXTRA!
Do humans have one, two or three sets of teeth?
Why don't babies have teeth?

How to be Brilliant at Living Things

Tooth decay

There are four different types of teeth in your mouth – incisors, canines, premolars and molars.

What you need: dental mirrors (if available); disinfectant; container; water; coloured pencils.

■ Show the different teeth by colouring in these diagrams. Use red for **incisors** (1 and 2), blue for **canines** (3), yellow for **premolars** (4 and 5) and green for **molars** (6 and 7).

Now, working in groups of four, think about what happens when you eat. Talk about what each type of tooth does.

■ Predict which tooth type decays the most.

■ Using a dental mirror, look at each other's teeth. Record fillings by showing a dot on the tooth like this.

■ Record missing teeth by showing a cross on the tooth like this.

■ Record the results for your group on this table.

Name	Number of teeth filled or missing			
	Incisors	Canines	Premolars	Molars
Total				

■ Which type of tooth shows the most decay? Was your prediction correct (see above)?

■ Why do the molars decay the most?

EXTRA!
Find out from books or CD-ROMs what causes tooth decay.

Favourite foods

People's tastes differ considerably, so their favourite foods will vary.
You need to eat a variety of foods each day to have a healthy diet.

What you need: graph paper; pencil; ruler.

■ Make a list of your ten favourite foods.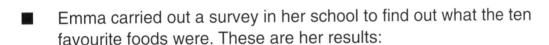

a How many of them are **sweet** foods?
b How many of them are **savoury** foods?
c Are any of these foods vegetables or fruit?
d Do you think you have a **healthy diet**?

■ Emma carried out a survey in her school to find out what the ten favourite foods were. These are her results:

Food	Number of children
crisps	9
baked beans	7
pizza	14
chips	11
bananas	1
fish fingers	3
chocolate	8
hamburgers	10
sausages	4
biscuits	6

a Which food was the children's favourite?
b Which food was bottom of the top ten?
c If you only ate the foods in the list above why do you think this would **not** be a healthy diet?
d Some of these foods are 'fast foods'. What is meant by 'fast food'?

■ Draw a bar chart on graph paper to show the results of this survey.

EXTRA!
Some people do not eat meat – they are called vegetarians. Why do people choose to be vegetarians? Do you think vegetarians have a healthy diet?

How to be Brilliant at Living Things

Healthy eating

It is important that we eat a nutritious balanced diet for our bodies to stay fit and healthy.

What you need: paper; pencil or pen; clipboard.

Which foods do you eat regularly? Which foods are good for your health and which do you think are not so good to eat? Are you and your friends getting a balance of different foods?

- Design a questionnaire to find out what children eat. First, work with a group of friends and make a list of different foods which you all eat regularly. Try to divide your list into types of food such as:

fatty foods:	chips, cheese, meat, crisps, butter
fruit:	apples, oranges, pears, mango, bananas
meat and fish:	lamb, beef, chicken, salmon, cod
starches:	pasta, bread, rice, potatoes
sugars:	sweets, jam, marmalade
vegetables:	carrots, peas, beans, cabbage

Your questionnaire could look like this:

Food	Once a day	More than once a day	Once a week	Never
Chips				
Cheese				

You could also include bread, potatoes, baked beans, crisps, eggs, salad, cakes, yoghurts, cereals, biscuits and sausages.

- Try out the questionnaire on other people in your group to see if it works. Then carry out your survey with at least ten other children.

- Does everyone eat fatty foods, fruit, meat and fish, starches, sugars and vegetables at least once a week?
  Does everyone have a **balanced diet**? Write down your findings.

EXTRA!
Write a menu for a healthy meal such as
a packed lunch or a hot evening meal.

How many times does your heart beat?

Your **heart** is the organ which pumps **blood** around your body.
It pumps blood to the top of your head and to the tips of your toes.

What you need: balloon; water; calculator.

■ Fill a balloon with water. **Outside or near a sink**, squeeze the
balloon gently.
   a   What happens?
   b   What does the balloon represent in your body?
   c   What does the water represent in your body?

Your heartbeat produces your pulse.
Use your fingers to feel your pulse in your wrist or your neck.

Count the beats you feel over 30 seconds.

Tip – Multiply this answer by 2 to find your pulse rate for one minute.

■ Calculate the following:

a)   My pulse rate and heartbeat   = _____   (beats each minute)

b)   My pulse rate for one hour   = _____   (beats per minute x 60)

c)   My pulse rate for one day   = _____

d)   My pulse rate for one month   = _____   (30 days)

e)   My pulse rate for one year   = _____

f)   My age is ____ years. So my heart has beaten at least _____ times.

g)   If someone is 80 years old, how many times has that person's heart
beaten?

■ Choose some of your relatives and calculate how many times their heart
has beaten.

EXTRA!
Using books or CD-ROMs find out information about the heart,
arteries and veins.

The heart, arteries and veins

The **heart** is the organ which pumps **blood** all around your body.
The diagram below shows the heart and the **blood vessels** carrying blood
to and from the heart.

**Plan of the body
showing the heart
and blood vessels**

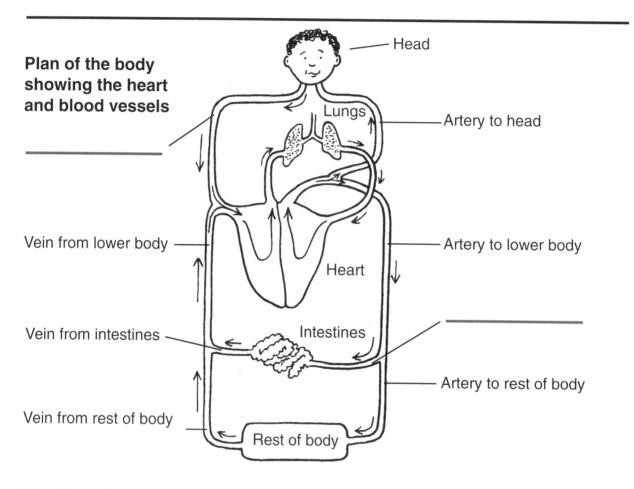

- Complete the two labels on the diagram above.

- What is the name of the blood vessels which carry blood **to** the heart?

- What is the name of the blood vessels which carry blood **from** the heart?

- How many chambers are there in the heart?

- Circle the two activities below which could cause heart disease.

smoking　　　　　　**much exercise**　　　　　**catching a cold**

eating fatty food　　　**not cleaning your teeth**　　　**eating fruit and vegetables**

EXTRA!
Imagine you have shrunk to the size of a pinhead and are in a mini-submarine
in your own blood system! Describe a journey from your heart and back again.

How to be Brilliant at Living Things

© Colin Hughes and Winnie Wade

Pulse rates

Your heartbeat causes your pulse rate. It is the rhythmical flow of the blood through the veins and arteries.

What you need: stop watch; calculator; graph paper.

Use your fingers to feel your pulse in your wrist or your neck. Ask your teacher to show you how to measure your pulse.

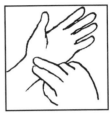 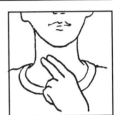

Measure your pulse rate for 30 seconds, then multiply by 2 to find your pulse rate for one minute.

Safety note: You should not carry out this exercise if you have been ill or if you have any breathing problems.

Work in pairs. Take your pulse before you begin, then take turns to run on the spot for two minutes. Make sure you stand away from furniture and other possible dangers. Measure your pulse rate again straight after the exercise, then after three minutes and after five minutes.

Fill in the table below.

Name	Pulse rate (number of beats per minute)			
	Before exercise	Immediately after exercise	After 3 minutes rest	After 5 minutes rest

■ Compare your pulse rate before and after exercise. How has it changed?

■ What happened to your pulse rate:
  a after three minutes rest? **b** after five minutes rest?

■ Why has the pulse rate increased after exercising?

■ On a separate sheet of graph paper, draw a bar chart to show how exercising and rest change your pulse rate.

EXTRA!
Find out more information about the heart, the heartbeat and pulse rate using books or CD-ROMs. Is the heartbeat the same rate as the pulse rate?

Pulsating data

Our pulse rate increases when we exercise and decreases after exercise.

■ What is pulse rate a measure of?

■ Explain how and where you would take your pulse rate.

A girl rests for two minutes, then exercises for one minute and rests for three minutes. Tick which graph shows her pulse rate.

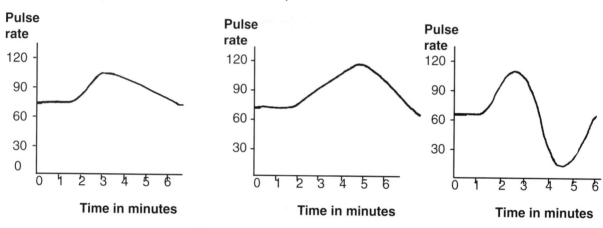

The graph below shows the pulse rate of a 'distance runner', before and after his race.

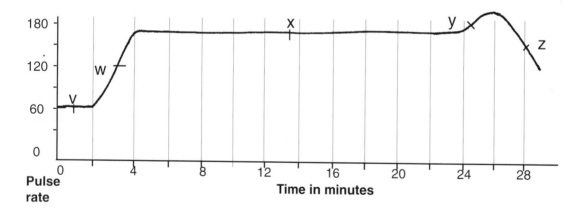

■ Give the pulse rate at each letter point on the graph and explain why it has increased or decreased.

The skeleton

Inside your body is a strong framework of bones called the **skeleton.** The 206 bones of the skeleton **support** the body. They also give **protection** to the organs in the body. Your bones, along with your muscles, allow you to **move**.

What you need: Resource sheet 1 (page 43); scissors; glue.

Here is half a skeleton.

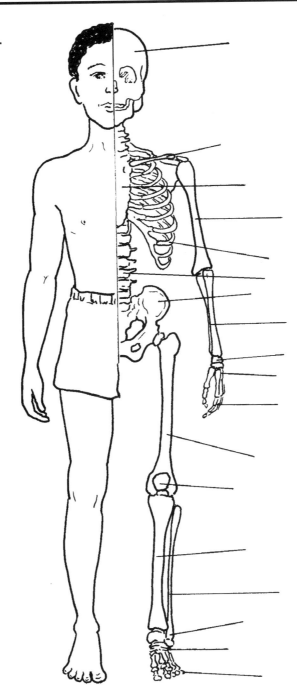

■ Use the labels from the resource sheet to identify the bones.

EXTRA!
Make a table showing the parts of the skeleton which protect the **brain,** the **heart** and the **lungs**, and the **reproductive organs**.

How to be Brilliant at Living Things

Comparing skeleton

The bones of the skeleton change throughout the lifespan of the animal. At birth the bones are soft, growing and strengthening to maturity.

These are skeletons of a baby, a grown-up human being and a rabbit.

■ Label some of the bones of thebaby's and rabbit's skeletons using the skeleton of the human to help you.

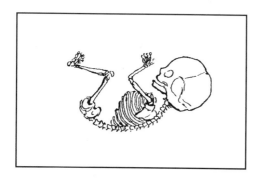

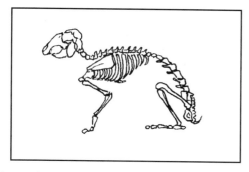

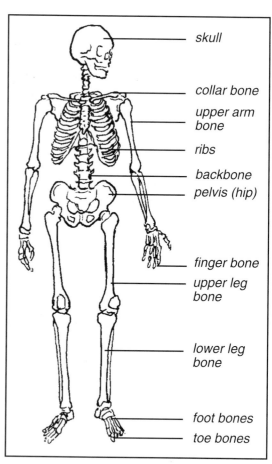

skull

collar bone

upper arm bone

ribs

backbone

pelvis (hip)

finger bone

upper leg bone

lower leg bone

foot bones

toe bones

■ How does the size of the baby's skull, ribs and leg bones compare with those of the adult?

■ What does this tell you about what happens to your skeleton as you grow older?

■ Does the rabbit have the same bones as a human (for example: skull, ribs and leg bones)?

■ Do the sizes of the rabbit and human bones vary?

EXTRA!
Look at diagrams of skeletons of other animals in books or CD-ROMs.
Do they have similar bones? Are some of the bones larger or smaller? Did the dinosaurs have ribs? Were their bones similar in pattern to ours?

How to be Brilliant at Living Things   © Colin Hughes and Winnie Wade

20   www.brilliantpublications.co.uk   This page may be photocopied by the purchasing institution only.

The human life cycle

The human life cycle shows the different stages that our bodies go through, changing from being a baby to being an adult. When we reproduce, the life cycle starts all over again.

Look closely at the diagram of the **human life cycle**.

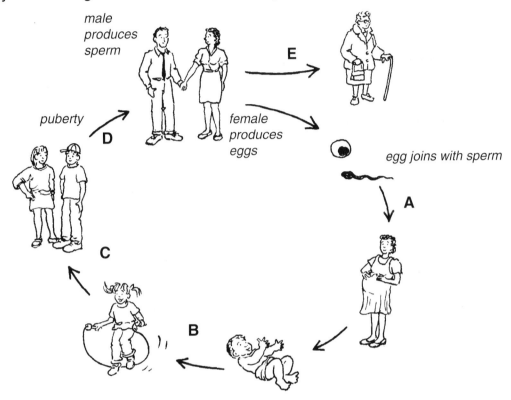

Describe the changes taking place at:

A _____

B _____

C _____

D _____

E _____

© Colin Hughes and Winnie Wade

This page may be photocopied by the purchasing institution only.

Smoking damages your health

Smoking can be harmful to your body. Every year thousands of people die from **smoking-related diseases**. There are now more and more places where smoking is banned.

What's in a cigarette?
Cigarettes are made from leaves of the **tobacco plant**. These leaves contain a substance called **nicotine**. Nicotine is a drug which people become addicted to – in other words they cannot do without it. Nicotine can make your blood thicker and move more slowly through your blood vessels.

Cigarettes contain **tar**, a thick, dark, sticky substance which settles on the lining of the lungs and damages them.

Carbon monoxide is a poisonous gas found in cigarette smoke. This gas makes it more difficult for the blood to carry oxygen around the body.

What effect can smoking have on your body?
Smoking can make you **short of breath** and more likely to have a bad cough or even a heart attack. You could get **lung cancer** or **cancer of the throat**. You are more likely to get other lung diseases such as **bronchitis**.

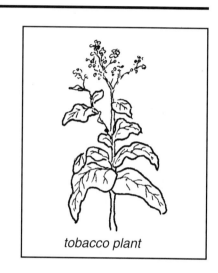

tobacco plant

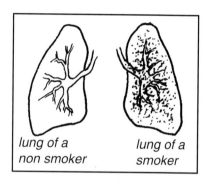

lung of a non smoker *lung of a smoker*

- Where have you seen a **no-smoking sign**?

- Think about ways in which people who smoke can be encouraged to give up smoking. How can we best explain to people that smoking is harmful to their health? Talk to your friends and make a list of things you could do to help people give up smoking.

- Design and produce a large poster which explains why smoking is bad for your health.

EXTRA!
Find out what effect smoking can have on people who do **not** smoke.

What affects plant growth?

Plants do not need pots to grow; they do not even need soil! So what do they require? Read on to find out …

Natasha and David grew a number of plants in ideal conditions. Then they moved the plants to different places in the school where the **temperature** varied. They did this to see how temperature affects plant growth. Here are their results:

Temperature	Height of plants (cm)
5° C	25 cm
15° C	30 cm
25° C	44 cm

■ At what temperature did the plants grow best?

■ At what temperature did the plants grow least well?

■ What other conditions did Natasha and David need to keep the same in order to make the test fair?

Natasha and David then looked at whether the amount of **water** affected how their plants grew. Here are their results:

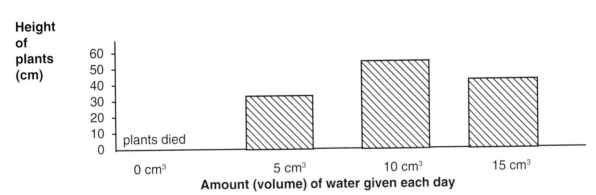

■ How much water was needed to make the plants grow tallest?

■ What height were the plants that were given 5 cm³ of water each day?

■ The plants that were given 15 cm³ of water each day did not grow as tall as those given 10 cm. Why was that?

EXTRA!
Design an investigation to see whether light affects plant growth.

Do plants need light?

What you need: piece of black plastic; piece of clear plastic sheeting; pegs, stones or pieces of wood to keep plastic in position; 2 plants or 2 trays of germinating cress seeds.

Lay the two pieces of plastic sheeting on healthy grass. Peg them down or put stones round the edge so that the plastic does not blow away.

Put a healthy plant or a tray of germinating seeds in a dark cupboard. At the same time, leave another plant or tray of germinating seeds in the light.

After five days, record your results on the table below. Use words such as *yellow, green, pale, healthy, unhealthy, spindly, elongated, normal.*

	Condition of grass	Condition of seedlings
Black plastic or dark cupboard		
Clear plastic or light place		

■ Do you think that light is needed for the healthy growth of plants?

■ Why did the grass under the black plastic sheeting and the plants grown in the dark look unhealthy?

■ What do most plants need to grow well? Circle the words below:

water   nutrients   warmth   worms   pots   light

EXTRA!
Design an investigation to see whether different amounts of light affect the growth of plants.

Leaves – the food factories

Plants are very important as they provide **food** for humans and other animals. We eat vegetables, fruit, corn and rice which are all parts of plants. Some of us eat meat from animals. Animals such as cattle eat grass; chickens eat corn. Their food, like ours, is produced by plants.

Plants need **light**, **water** and **warmth** in order to have healthy roots, stems and leaves and to grow well.

The **roots** take up water into the plant. They also take up nutrients which are important for the health of the root, stem, leaves and flowers.

The **stem** carries the water to the rest of the plant. The **leaves** are produced on the stem.

Plants do not have mouths so they do not **eat** food. They have to **make** their own food. The leaves are very important as they produce food for the plant so that it can grow. The leaves must have light to produce this food. If a plant is put in a dark place no food is produced and the plant becomes yellow and spindly.

- Which part of the plant makes food?

- What does this part of the plant need to produce the food?

- Apart from light, what else do plants need for healthy growth?

- What does the stem do?

- If you put a plant in a dark place for a long time, would it die? Give reasons for your answer.

EXTRA!
Take three plants of the same height. Leave one as it is. Remove most of the leaves from the second and all of the leaves from the last one. Which plant grows the best?

How to be Brilliant at Living Things

From root to leaf and flower

Roots take water and nutrients from the soil, which is then transported up the stem to the leaves and the flowers.

What you need: celery stem with leaves; an uprooted plant such as groundsel; 2 containers; food dye (red); knife.

How does water get to the top of the tallest tree? Let's find out by using celery and groundsel rather than a tree!

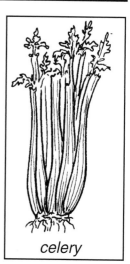

groundsel

celery

■ Place 2 cm³ of water in each container. Add 10 drops of food dye (be careful not to spill it or it will stain your hands and clothes). Wash the groundsel under the tap to remove the dirt. Place the celery and groundsel in the containers.

■ Predict what you think will happen to the celery and the groundsel.

■ Every five minutes record how far the dye has travelled up the plants. Record your results in the table.

Time (minutes)	Height of dye	
	celery	groundsel
5		
10		
15		
20		
25		
30		
35		
40		

■ When the dye has reached the top of the plants cut through them at different places. Draw pictures of what you see.

■ Complete the passage below using the following words:
**flowers   leaves   nutrients
roots   soil   stem**

Water and _____ are transported from the _____ to the

_____. From the roots they pass through the _____ of the

plant. From there they travel to the _____ and the_____ of

the plant.

EXTRA!
Draw a bar chart or graph of your results.

How to be Brilliant at Living Things

© Colin Hughes and Winnie Wade

Buttercup

The buttercup is a common meadow plant with yellow cup-shaped flowers.

What you need: magnifiers; buttercup flowers; wide sticky tape.

Here is a diagram of the inside of half of a buttercup flower.

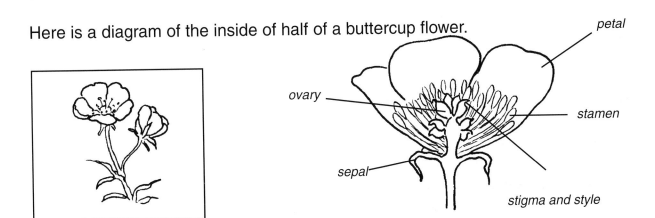

- ■ Work with a friend to learn about the parts of the flower.

- ■ Collect a buttercup and a magnifier. Look carefully at the parts of the flower and compare them with the picture above.

- ■ Carefully remove the parts of the flower and lay them in the correct spaces below. Cover and stick the parts down with sticky tape.

Sepals	Petals
Stamens (male part of the flower)	Stigma, style and ovary (female part of the flower)

EXTRA!
Collect some different flowers and look at them carefully. See if you can identify all the parts of these flowers.

How to be Brilliant at Living Things

Life cycle of a plant

When a plant grows, it goes through a number of different **stages**. These stages take place in a certain order and are repeated over the years. The **life cycle of a plant** shows how it grows and reproduces.

What you need: Resource sheet 2 (page 44); scissors; glue.

The pictures on the resource sheet show the stages in the life cycle of a plant, but they are in the **wrong** order! Can you put them in the **right** order? Cut out the pictures and arrange them in a circle. The seed has been completed for you. Then stick the correct label near to each picture. Draw arrows to show the direction of the life cycle.

seed

EXTRA!
Look at other plants in the school and outside. Match them with the stages of the plant life cycle on your sheet.

Plant and human life cycles –
how are they alike?

A life cycle shows how a new individual is formed, how it grows and in turn produces other new individuals. It is called a cycle because it keeps on going around.

Life cycle of a flowering plant   **Life cycle of a human being**

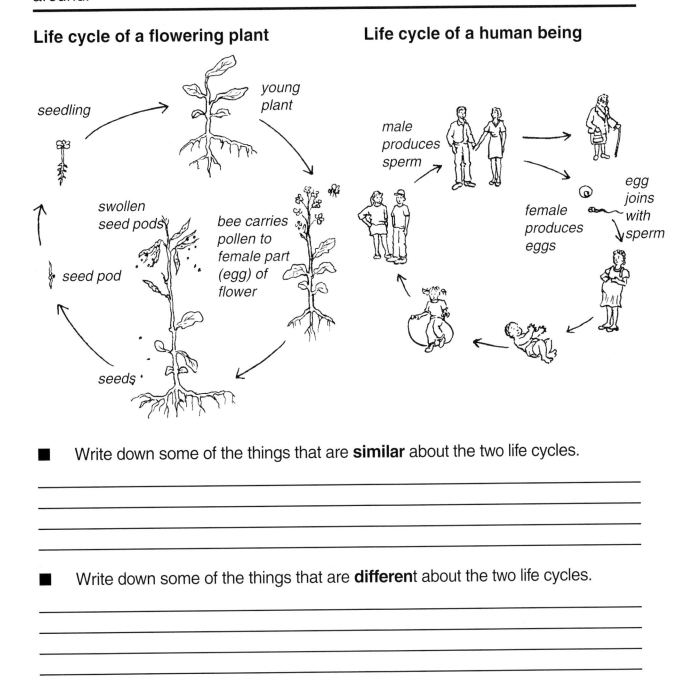

- Write down some of the things that are **similar** about the two life cycles.

_____

_____

_____

- Write down some of the things that are **different** about the two life cycles.

_____

_____

_____

_____

EXTRA!
Find out the differences between **annual**, **biennial** and **perennial** plants.

How to be Brilliant at Living Things

Scattering seeds

Seeds can be scattered away from the parent plant in a number of ways. Some are carried by animals and birds; others float on water; some are blown by the wind; others are scattered by the plant's own explosive mechanism.

What you need: A collection of different types of seeds – sycamore or ash seeds; thistle, dandelion or groundsel seedheads; poppy seed-cases; pea-pods or broom-pods; goosegrass; seeds of juicy fruits; acorns; magnifiers; Resource sheet 3 (page 45) if the seeds and fruits above are not available.

■ Work with a friend. Look closely at the seeds (or the pictures on resource sheet). Try to name them.

■ How is each seed carried away from the parent plant? Look for clues to help you. Wings? Sticky? Hooks to cling to your clothes? Pods which will dry out and explode? Record your results in the table below.

Drawing	How is it scattered?	Seed name

EXTRA!
Why is it important that seeds are scattered
away from the parent plant?

Plant test

Test your knowledge and understanding of plants.

Write **T** for **True** or **F** for **False** in the boxes below.

The roots of a plant take in water. ☐

The roots of a plant produce seeds. ☐

The roots of a plant produce leaves. ☐

The roots anchor the plant in the soil. ☐

The roots of a plant are green. ☐

The roots of a plant make food for the plant. ☐

The roots of a plant take in nutrients. ☐

The stem of a plant produces leaves. ☐

The stem transports water and nutrients. ☐

Plants, like animals, take in food. ☐

The leaves of a plant are green. ☐

The leaves make food for the plant. ☐

Food production in the leaf requires light. ☐

All plants need light and water to grow. ☐

All plants need warm conditions to grow. ☐

The male part of the flower is called the sepal. ☐

The stamen produces pollen. ☐

Pollination is when a pollen grain joins with an egg. ☐

Fertilization takes place in the ovary. ☐

In the life cycle of a plant, fertilization takes place
before pollination. ☐

Germination is when a seed starts to grow. ☐

Sycamore seeds are dispersed by animals. ☐

Dandelion seeds are dispersed by the wind. ☐

Identifying buds

As plants grow and develop they produce buds. Buds are part of the plant's life cycle, developing into either the flowers or the leaves.

The key below describes the buds on twigs of trees.

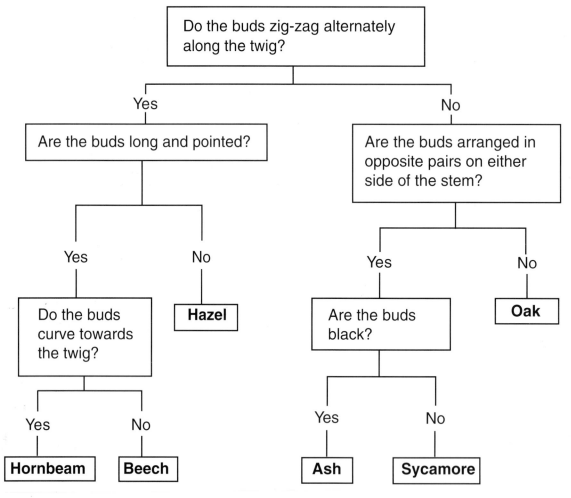

Use the key to identify the tree buds and twigs below.

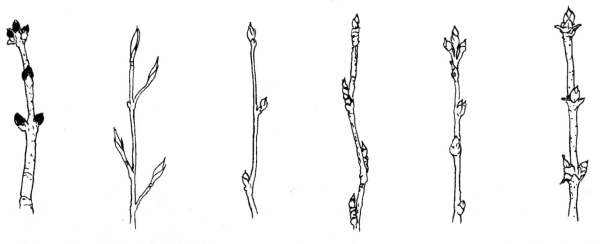

How to be Brilliant at Living Things

32   www.brilliantpublications.co.uk

© Colin Hughes and Winnie Wade

This page may be photocopied by the purchasing institution only.

Which leaf?

What you need: a collection of leaves to include: ash, beech, horse chestnut, holly, lime, oak, sycamore, willow. If leaves are not available, use Resource sheet 4 (page 46).

Use the key below to identify the leaves in your collection (or on the resource sheet).

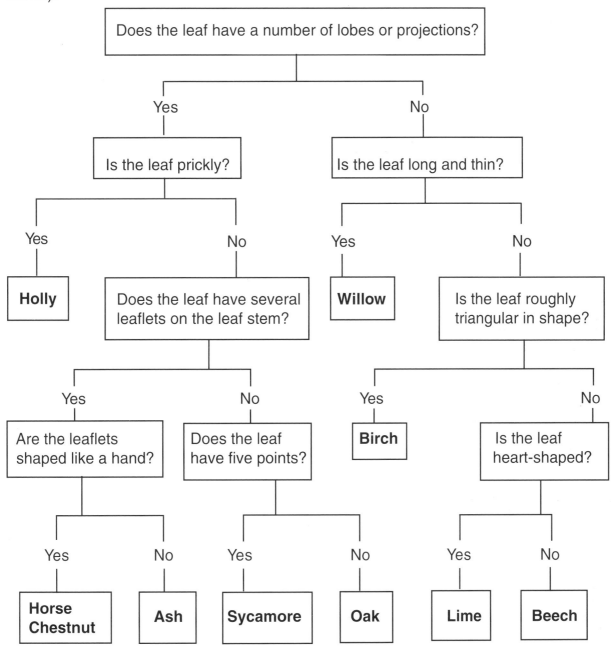

EXTRA!
Collect some more leaves and make a key of your own with different features.

© Colin Hughes and Winnie Wade

How to be Brilliant at Living Things

This page may be photocopied by the purchasing institution only.

Leaf key

We can use keys to help us to identify living things. It is quite easy to design a simple key. It means thinking of some good questions and putting things into groups or sets.

What you need: a number of named leaves of different types including: sycamore, horse chestnut, beech, ash, oak and pine.

- Look at your leaves. Make a list of the similarities and differences between them.

- Ask yourself **yes** or **no** questions about the leaves until you have sorted and identified every one. (Look at some other keys, such as the one for identifying buds (page 32) to give you some ideas.)

- Write your leaf key down. The first question has been done for you, or you could write one of your own.

- Try out your key with a friend.

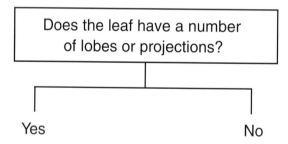

```
┌──────────────────────────────────┐
│  Does the leaf have a number     │
│  of lobes or projections?        │
└──────────────────────────────────┘
          │
   ┌──────┴──────────────┐
   │                     │
  Yes                    No
```

Classifying animals with backbones

Animals with backbones are called **vertebrates**. The bones in the spine or back-bone are called **vertebrae**. These animals include **amphibians**, **birds**, **fish**, **mammals** and **reptiles**.

There are many different vertebrates so it is helpful to sub-divide them into smaller groups:

- **Amphibians** have smooth, moist skin
- **Birds** have a beak and feathers
- **Fish** have fins and scales

- **Mammals** have hair and give birth to live young
- **Reptiles** have dry scales but do not have fins

■ Identify these animals and fill in the table below.

Animal	Vertebrate group	Reason for choice
cod		
eel		
frog		
human		
lizard		
robin		
snake		
swan		
tortoise		
whale		

■ Using books or CD-ROMs find out which of the five groups have a constant body temperature and which change their body temperature as the air temperature changes.

EXTRA!
Design a key to identify the five groups of vertebrates.

How to be Brilliant at Living Things

Some like the light, some like the dark!

Different animals and plants are found in different **habitats**. A habitat is the natural home or environment of the animal or plant. Some creatures like the dark and they spend most of their life either active at night or hidden under bricks, pieces of wood or piles of leaves, while other creatures prefer the light.

What you need: large ruler or stick; collecting trays or blanket; pooters or plastic trays; plastic dishes; trowel; plastic bags; book for identifying insects and other minibeasts; hand lenses.

Hedgerow (light)

Shake the twigs and branches of a hedge using a large ruler or stick. Hold a collecting tray or blanket underneath. Use a pooter to collect any animals.

Leaf litter (dark)

Dig down into a pile of dead leaves. Take the leaves back to the classroom in a tray or plastic bag. Sort through the leaves in a tray looking for animals.

■ Using a hand lens, look closely at the animals you find. Try to identify them using keys or pictures in books. Draw pictures of the animals in the spaces below.

Hedgerow (light)	Leaf litter (dark)

■ Did you find the same animals in both the light and dark habitats?

■ What have you found out about where animals live?

Remember – return the animals to their original habitats when you have finished working with them.

EXTRA!
Put a log or some large stones in your school grounds. After several days look underneath. Are the animals similar to those found in the leaf litter or the hedgerow?

Light or shade? Dry or damp?

Some animals like shade, other animals like to live in the light. Some animals prefer to live in the dry, while other animals prefer the damp.

What you need: margarine tub; a tray with sides or tub in which to carry out an investigation; a spoon; dark coloured paper; filter paper; sticky tape; ten woodlice; stopwatch.

Remember to be careful with the woodlice at all times and to return them to where you found them when you have finished with them.

- Use a tray or tub to design an investigation to find out which four conditions the woodlice prefer.

- How will you change the tray or tub to find out which one they prefer:
  a   light or shade?
  b   dry or damp?

- When will you decide whether they prefer one condition as opposed to another?
  after   **a**   10 seconds
  b   1 minute
  c   10 minutes
  d   1 hour

- Plan out your investigation in steps.

- Make a table to record your results.

- Which condition did the woodlice prefer?

EXTRA!
Find out where woodlice live outside. How do these conditions compare with the results of your investigations?

How to be Brilliant at Living Things

Different habitats

An area where plants and animals live is called a **habitat**.
Different habitats have different plants and animals living in them.

What you need: Resource sheet 5 (page 47); scissors; glue.

■ Cut out the pictures on the resource sheet. Stick them on the correct habitat below.

Freshwater habitat

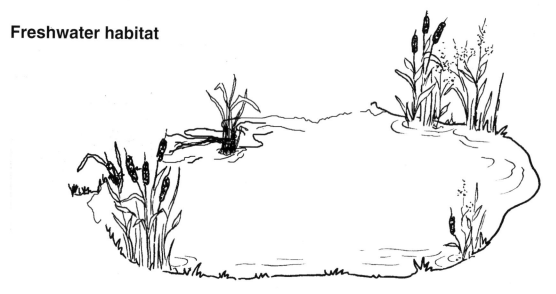

Woodland habitat

■ Label the plants and animals. Use the words on the resource sheet to help you.

EXTRA!
Draw the plants and animals you would probably find in:
a **desert**; a **polar habitat**.

Why are they found there?

Animals and plants live in particular **habitats** or environments because the conditions suit them. They are **adapted** to the conditions. For example, a fish has fins to help it to swim or a desert rat is a similar colour to the sand so that it is camouflaged.

■ Write down how each animal or plant is suited (adapted) to its habitat or environment. The labels should give you a clue!

Freshwater habitat

gills, fins tail,

leaves on surface

rooted in bottom

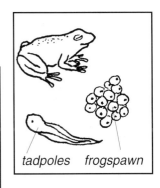

tadpoles   frogspawn

Desert habitat

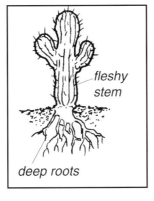

fleshy stem

deep roots

hole/cave at night

basks in sun during day

large feet, big eyelids

How to be Brilliant at Living Things

Food chains – predators and prey

A **food chain** shows how living things in a certain habitat are linked by what they eat.
Producers are always plants; they produce food for animals from the energy from the sun.
Predators are animals which eat or prey on other animals. Animals which are eaten by other
animals are called **prey**. Any living thing which eats another living thing is called a **consumer**.

What you need: Resource sheet 6 (page 48); scissors, glue.

■   Use the pictures on the resource sheet to make three food chains.

Food chain 1

Food chain 2

Food chain 3

■   Write down the animals and plants that you have chosen for your food chain
      in the table below.

Producer	Prey	Predator or consumer

Micro-organisms

There are many millions of micro-organisms. We can divide them into those that are harmful and those that are useful. **Harmful** micro-organisms can make food decay or make us ill. **Useful** micro-organisms can help make medicines such as antibiotics and can be used to make some foods.

■ Look at the pictures and decide whether you think useful or harmful micro-organisms have had a part to play. Write **useful** or **harmful** underneath each picture.

■ Draw some pictures of your own to show how micro-organisms can be harmful or useful. You can find several examples in the garden.

EXTRA!
Find out what these words mean: **microbe**, **virus**, **bacterium** and **germ**.

How to be Brilliant at Living Things

Mouldy food

Micro-organisms are present in the air. If we leave food out in the open for too long micro-organisms can infect the food, resulting in a growth of bacteria or mould. Moulds and bacteria cause the food to go rotten.

What you need: slices of bread (moistened if necessary); plastic bags with ties or clear plastic containers with lids.

Keeping our hands and kitchen surfaces very clean when preparing and serving food reduces the risk of contamination.	**We can become ill with 'food poisoning' if we eat food that has been infected with some micro-organisms.**

■ Write down some ideas for preventing food from going mouldy or becoming infected with harmful micro-organisms.

■ Design an investigation to find out which conditions cause bread to go mouldy most quickly.

Here are some ideas to help you:
◆ What will you keep the same during your investigation to make it a fair test?
◆ Which **one** thing will you change?
◆ Look closely at the slices of bread every day and write down and draw on the back of this sheet any changes that you can see. FOR SAFETY REASONS, KEEP THE BREAD COVERED AT ALL TIMES. You might have to wait several days before anything happens. Predict in which condition the bread will turn mouldy first.
◆ Why will this slice of bread start turning mouldy first?

■ Did any mould grow on any of the slices of bread? If so, which of the slices did this happen to first?

■ Can you explain why mould grows more quickly in some conditions than others?

EXTRA!
Make a poster to show some simple ways of preventing food being contaminated by micro-organisms.

The skeleton

The skeleton is the framework of bones that support your body.

upper arm bone	breastbone
lower arm bones	backbone (spine)
ribs	wrist bones
ankle bones	finger bones
collar bone	skull
toe bones	hip bone (pelvis)
shin bones	foot bones
calf bone	thigh bone
hand bones	kneecap

Enlarge the diagram from page 19. Cut out the labels above and use them to identify the bones in the skeleton. Stick them next to the correct part in the picture.

How to be Brilliant at Living Things

www.brilliantpublications.co.uk

Life cycle of a plant

The life cycle of a plant shows the stages of a plant's life span, how it grows and reproduces.

Resource sheet 2

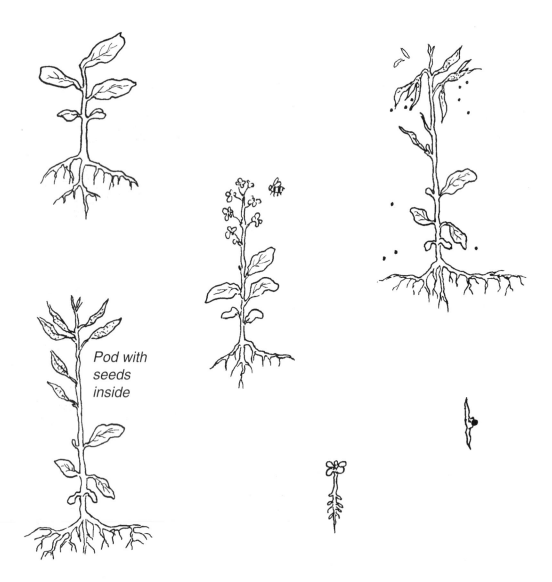

Pod with seeds inside

young plant	seed and fruit production	seed dispersal
germination	flowering and pollination	seedling

Scattering seeds

There are a number of ways in which plant seeds or pods are dispersed or scattered from the parent plant.

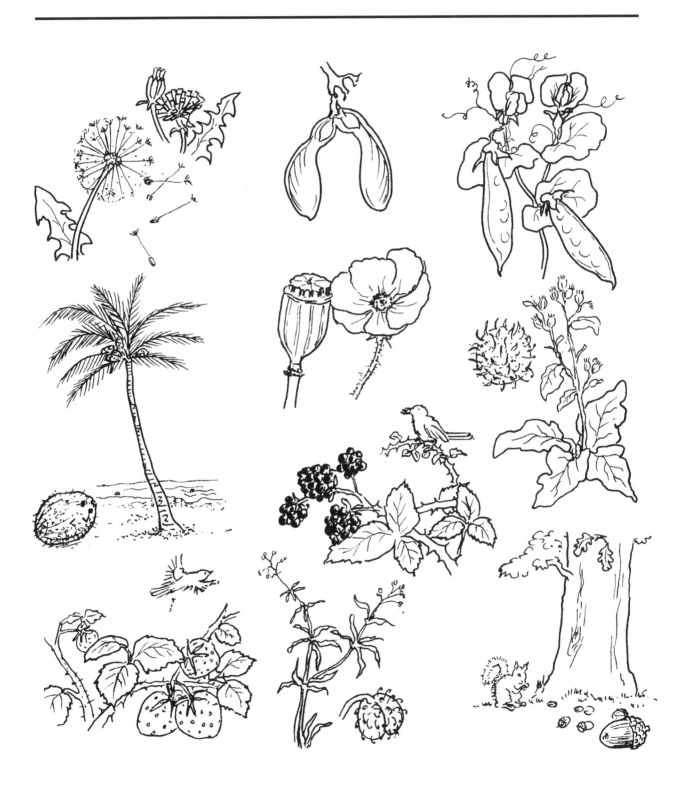

Which leaf?

leaflet

leaf
5–7 leaflets
4–20 cm long

projection

leaf 7–10 cm long

lobe

leaf 5–9 cm long

each leaflet 4-9 cm long

leaf 3–6 cm long

leaf 5–25 cm long

leaf 3–4 cm long

leaf 5–12 cm long

leaf 4–8 cm long

Identify the leaves using the key chart shown on page 33.

Different habitats

A **habitat** is the natural home or environment of a plant or animal. It is the surroundings where plants or animals prefer to live.

Resource sheet 5

mallard	squirrel	dragonfly	blue tit	frog
grass snake	bulrush	primrose	trout	newt
ash	fox	waterlily	bluebell	oak	owl	willow

Cut out the animals and plants above and stick them in the correct habitat on page 38. Use the labels above to help identify each plant or animal.

How to be Brilliant at Living Things

Food chains; predators and prey

There has to be an order to the food chain for every plant and animal to survive. As the smaller creatures are being eaten on a more regular basis, you will find that overall they are more abundant.

Resource sheet 6

■   Cut out the pictures of the plants and animals below. Read the 'what eats what' clues at the bottom of the page. Make three food chains on page 40 and draw arrows showing 'what eats what', for example: lettuce ✿ slug.

sparrowhawk

lettuce

owl

fox

thrush

rabbit

slug

dormouse

corn

grass

Sparrowhawks eat thrushes.
Rabbits eat grass.
Thrushes eat slugs.
Dormice eat grain.

Rabbits are eaten by foxes.
Slugs eat lettuce.
Dormice are eaten by owls.

■   Which **type** of living thing does each food chain start with?

■   Make a list of all the predators and another list of the prey that appear in the food chains above.
